A SPACE ELEVATOR ON THE MOON

Book 5 of the Space Elevator 2020 series

Linda Phillips

CONTENTS

A SPACE ELEVATOR ON THE MOON

Book 5 in the Space Elevator 2020 series

We're returning to the Moon... using rockets. Yet setting up a space elevator on the Moon is easier than doing the same on Earth, thanks to the low gravity of the Moon. A ribbon doesn't need to orbit the Moon, unlike Earth. The pull of Earth's gravity means we can hang a ribbon "down" towards Earth, a ribbon that could stretch halfway to Earth. So, when will it happen? Will it be the USA that constructs it, or China?

The concept of a Space Elevator started over a century ago but started to move from science fiction to reality in the 1990s with the promising advent of a super strong material, carbon nanotubes, that theoretically had the strength to build a ribbon into space.

NASA started investigating the concept which led to the publication of the first detailed analysis of whether it was possible, written by Dr Brad Edwards, in a NIAC paper.

Early excitement of an immediate breakthrough, maybe even constructing one by 2010, gave way to the reality of producing carbon nanotubes of the required strength, outside the laboratory, in commercial quantities. As I write, in 2020, we are still awaiting this breakthrough and the project has moved back to the late 20s, maybe the 2030s or 2040s.

But it will happen, and when it does, it will replace rockets as the easier and cheaper way to leave Earth, opening up space travel the way airplanes opened up world travel.

This series examines the technical aspects and how it will be deployed.

In book five we set out the current plans for a return to the Moon, then look at how the use of a space elevator on the Moon changes the dynamics and makes Moon colonies a practical advance.

The related book by Linda Phillips, "Back to the Moon", details the existing rocket based activity to return to the Moon and what will happen in the period prior to the deployment of a space elevator.

DEVELOPING THE CONCEPT

In 2003 Philip Ragan came up with the concept of a space elevator on the Moon and set out the mechanics of it. The concept was included in the 2006 book "Leaving the Planet by Space Elevator", by Ragan and Dr Brad Edwards, NASA physicist who did the original analysis of the space elevator concept in the 90s. With the passing of her partner Philip, Linda Phillips and Dr Edwards reissued the book under their names, with minor updates.

In 2020 Phillips produced the Space Elevator series with a complete update reflecting all the work on the concept that has occurred since and they are available on Kindle.

FREQUENTLY ASKED QUESTIONS

Building a space elevator on the Moon brings different challenges and opportunities to one on Earth. These are covered in detail in the book. Here we summarize key factors.

Is it easier to erect a space elevator on the Moon than the Earth?

Yes, as gravity, at the surface, is only 16.5% that of Earth, the apparent weight of an elevator ribbon is 16.5% less, so the ribbon doesn't need to be as strong. So, we may have the technology to build an elevator on the Moon before we build one on Earth.

Can we connect Earth and Moon with a ribbon?

No, great though that would be. The Moon doesn't stay in one place, relative to a point on the surface of the Earth. Go outside and look up in the sky. Can you see the Moon? If so, in two weeks' time you won't be able to see it, as it will be on the other side of the Earth. The Moon would have to be "geostationary" to perform that feat.

But we can connect a ribbon from Earth to satellites in space?

Yes, because most of our satellites are positioned some 36,500 km from Earth in a zone called geostationary orbit (GEO). Put a satellite there and it orbits the Earth once each Earth day of 24 hours approx., creating the illusion the satellite doesn't move when you look at it from Earth. So when building a space elevator from Earth (see previous books in this series for the detail), the ribbon will connect to a space station at GEO, which we call the GEO Station, and the ribbon continues away from Earth, for another 65,000 km or so, where a terminus station hangs on to the end of it, whimsically called the Far Far Away Station. Rockets can be launched into space from this station, getting the benefit of the orbital speed of the station. At first, the Moon will be a prime target for rockets, leaving this station to the Moon or to the space elevator on the Moon.

If the space elevator from Earth is 100,000 km long, how close to the Moon does it get?

On average, the Earth-Moon distance is 400,000 km, so the space elevator from Earth gets a quarter of the way there. This is better than it sounds. The most expensive part of the journey is the first 10,000 km leaving Earth, as escaping the Earth gravity is the hard part. Once you are 100,000 km from Earth, gravity seems almost non-existent in practical terms so a transfer to the Moon is simple.

Would a Moon space elevator stay up with centrifugal force, like the Earth one?

No, different mechanics involved. The Earth rotates fast enough that a ribbon extending past GEO can stay in place thanks to the orbital velocity. But, while the Earth rotates once a day or so, the Moon takes about 28 Earth days to rotate once. As it also orbits the Earth roughly once every 28 days, the same face of the Moon is always pointed towards Earth, tidally locked.

From Earth, a rotation once an Earth day is fast enough that we can use centrifugal force to keep the ribbon in place. But the Moon doesn't rotate fast enough to do the same trick.

Fortunately, there is an easier way of keeping a Moon ribbon in place. Think of a ribbon, fixed to the Moon, hanging down towards Earth. As the effect of Earth gravity is strong relative to Moon gravity, it's like a stone on a string, hanging down towards Earth. Earth's gravity pulls on it, dragging it towards Earth and that is how the ribbon stays in place on the Moon.

So, in summary, an Earth ribbon has to use centrifugal force to stay in place, while a Moon ribbon is pulled by gravity towards Earth and stays in place, just like, if you hold a string with a stone on the end of it, it hangs down towards Earth.

HOW LONG WILL THE MOON RIBBON BE?

The Earth-Moon gap is about 400,000 km (it does vary a bit during its monthly cycle and other reasons). The Earth space elevator stretches for 100,000 km, a quarter of the way there.

The Moon ribbon would ideally be 150,000 km to 200,000 km long, in order to be captured by Earth gravity and hang there. If it is 200,000 km long, then picture this: once a day or so, as the Earth rotates, the Earth ribbon swings past the Moon and at its closest approach, there is only a gap of 100,000 km between them.

The Earth ribbon takes you a quarter of the way to the Moon. You use a rocket to travel another 100,000 km to link up to the Moon ribbon, half-way to the Moon, then the Moon ribbon takes you the remaining 200,000 km to the Moon.

Can the Moon ribbon be a different length?

Now we get into the tensile strength of materials. For the ribbon, it is in tension (the opposite of compression), stretched by the gravity of the Moon and Earth respectively, and affected by the amount of mass of cars traveling on the ribbon.

A shorter Moon ribbon, say 150,000 km, has tensile forces low enough it could possibly be made of a strong metal like Tungsten, which means, in theory, we could build it today. But once we get strong enough carbon nanotubes in production, they would enable us to stretch it to 200,000 km. We have not yet assessed just how strong commercially produced carbon nanotubes would become, but the next goal would be to produce a ribbon stretching

300,000 km from the Moon.

A 300,000 km Moon ribbon would, in theory, pass immediately by the Earth ribbon, once a day, or at least get in close proximity, allowing for the many variations in the position of the Moon relative to Earth. In this scenario, we can imagine, transferring from one ribbon to the other would be a process akin to hopping off one and jumping on to the other. In practice, there is more to it: the difference in velocity still calls for a rocket propelled vehicle to make the necessary speed adjustments; the journey length could be anywhere from a nominal kilometer to many thousands of kilometers; it may involve changes in the orbital plane requiring the use of propulsion.

The mechanics of the transfer can be managed, but in principle it's not so different from doing a transfer of 100,000 km distance. It would be a kind of neat solution, though.

What if we could extend the Moon ribbon to a length of 400,000 km, say just grazing the Earth atmosphere?

Doesn't help. The effect of Earth gravity, that close to Earth, is the same as for an Earth ribbon. The pull of gravity could threaten and break the ribbon, done that way. In any case, the velocity of such a ribbon, rushing past Earth, would make it difficult for a rocket to leave Earth and dock with it.

So, over a period of approximately 25 hours, the Moon will have circled the Earth (or, to be strictly correct, the Earth will have turned to the point where the Moon is in the same position again in the sky). So, if we dropped a ribbon down from the Moon to the Earth, how fast would the ribbon move across the surface of the Earth?

The answer is a very fast 1,600 kilometers per hour or 1,000 miles per hour, about twice the speed of a jumbo jet at cruising speed!

A ribbon is only useful at the Earth end if you can climb on board! A ribbon running by you at that speed is totally useless: it would be worse than trying to jump on board a jumbo jet passing by at full throttle.

When even a jumbo jet couldn't fly fast enough to connect with the ribbon, then we know that it won't work!

It's not just the speed: drag a ribbon through the atmosphere at 1,670 kph and the friction will heat it up and it will eventually burn up.

Lastly, the cost. Recall the most expensive part of getting into space, is the first 10,000 getting away from Earth. In this scenario, we still use rockets for the most expensive part, before using a low-cost space elevator to reach the Moon.

The most effective scenario remains one of having an Earth ribbon taking cars to GEO and the Far Far Away Station, using lightweight rockets to leave that station and dock with the Moon ribbon, then descending the Moon ribbon to the lunar surface.

WHERE WILL THE MOON BASE BE?

In book 4 in the series, "Space Elevator Earth Ports", we listed the best locations on Earth to locate a space elevator, most of them being islands in the Pacific Ocean.

Finding a location on the Moon is easier than on Earth. There are no countries, cities or populations to take into account, for starters.

On Earth we prefer a floating terminal base so it can be moved around, for several reasons. When setting up, ribbons will drop down to the equator and then dragged to the base location north or south of the equator. With the excessive amount of satellites and debris in LEO (low Earth orbit), including the thousands of small comms satellites being launched, the ribbon needs moving to avoid collisions at times.

The lunar ribbon has different dynamics. Since the ribbon hangs down towards Earth and (so far) the Moon doesn't have lots of satellites in orbit around it, the lunar ribbon can have a fixed location. In practice we wouldn't want to move a Lunar ribbon as we don't have an ocean handy, which means trying to transport it across the dusty and unforgiving lunar surface.

In our original 2006 book, Leaving the Planet by Space Elevator, we suggested the crater of Alpetragius as the base for the Moon space elevator. Since then, the availability of high definition mapping of the lunar surface has enabled an examination of potential sites in more detail. This is discussed in more detail later in the book, but for now, the focus is on the similarly named Albategnius crater. It is near the center of the Moon, as viewed from Earth. At 129 km in diameter it has a number of useful smaller craters within it, and, most important, it has a mountain in the middle of it, 1.5 km in height, with a small crater on top.

Putting a colony building on the surface of the Moon is a challenge (see later). A better alternative, once a ribbon is connected to the central mountain, is to tunnel into the mountain and develop a colony underground, where it is protected from heat, cold, radiation and meteorites.

As a ribbon drops to the lunar surface during the initial construction, it will need steering towards Albategnius, which is 360 km from the center point, but the low lunar gravity means this can easily be achieved using rockets. Once connected and secured to the lunar surface, the ribbon can drop towards Earth and take up its permanent position.

What else can we do besides connect a ribbon to Albategnius crater?

The ribbon is there for transport connections onto and off of the Moon. But carbon nanotube is a marvelous material. The next problem for a colony is the absence of air and air pressure on the surface, so spacesuits are needed to get around.

But, using carbon nanotubes, we can create a net-like fabric, like a spider's web or a cotton sheet. Build it big enough to cover the entire crater, like an umbrella. Raise it in the center, held up by the ribbon, and just like a circus tent, we can cover the entire crater with a roof! Line the carbon nanotube roof with a solid membrane material. Then we can pump air into the crater until we have an air-filled crater.

Now we're talking! Instead of a hostile vacuum surface, we can create a liveable, air-filled space, in this case 129 km across and this can be used to create the first real colony on the Moon! Add soil and we can grow crops for the colony, as well as providing parks and recreation space.

In comparison, the colonies currently proposed, based on rocket technology, are temporary camps, nowhere near the real thing.

CAN WE DEPLOY MORE RIBBONS ON THE MOON?

Yes, we can. If we think of the first ribbon as the central spine to the Moon, we can add more ribbons. From a suitable point in space a few thousand kilometers up the ribbon, we can create a station, and run more ribbons to almost anywhere on the Moon.

Ribbons to the poles will be a first priority as that is where the water is. Due to the curvature of the surface we can't run a ribbon all the way to the lunar horizon, but since the Moon is so small, we can get close. Then it is a matter of creating roads or railways to connect to the water sources or anywhere else we want to be.

Imagine a dozen such ribbons descending, like the spokes of an umbrella, to various points on the lunar surface. With this transport network, we are capable of colonizing the Moon in a practical way.

What's the best way to travel around the Moon?

The regolith on the lunar surface is nasty stuff. It consists of very fine dust, radioactive due to the lack of the Van Allen Belt that protects Earth and has an electrical charge which means it sticks to everything. As the Apollo astronauts discovered, the electrical charge on the dust means it gets everywhere.

As a result, travel on the lunar surface is a problem. But with a network of ribbons as described above, it will be faster and safer to use them like a rail network: ascend to the main station and descend down another ribbon to your destination, avoiding travel on the actual surface as much as possible.

The other useful advance will be to take poly-tunnels such as we build on Earth, only larger, and stretch them like roads between destinations. A poly-tunnel can be filled with air, and it excludes the outside environment, keeping that nasty dust out. So, we can use such tunnels for travel between colonies, mines, etc.

What about the far side of the Moon?

In principle, we can build a space elevator on the far side of the Moon, the side which always faces away from Earth.

There is interest in building colonies on the far side. China has landed an explorer on the far side, Chang'e 4, and more exploration is likely to follow. It is attractive to astronomers, avoiding the radio noise emanating from Earth, which makes it a great place to build telescopes of all kinds.

On the near side of the Moon, a space elevator ribbon is kept in place by dropping down towards Earth, like a stone on the end of a string.

Building a space elevator on the far side utilizes the opposite effect. The ribbon needs to be much longer, passing through the Lagrange Point, so that the small amount of centrifugal force generated by the Moon orbiting the Earth is sufficient to keep the ribbon in place. The mechanics are more delicate than for the near side, but theoretically possible.

IT'S ALL ABOUT GETTING OFF OF EARTH

The delta-v required to leave Earth for the ISS in LEO is about 9.4 km/s or 9400 m/s. The delta-v required to move between space elevators, or to orbit the Moon or to adjust orbits around the Moon, to reach the proposed Gateway, is in the range of 200 m/s to 900 m/s.

This shows clearly: the biggest problem to solve is getting off of Earth. It costs 10 to 50 times more to get off of Earth than to maneuver in space around the Moon.

The above FAQs cover key facts about building a ribbon to the Moon. In the next chapters they are addressed in detail.

Jumping off into space

There are many practical reasons for humans to travel a Space Elevator without leaving it. There will be maintenance crew, vehicle crew, and space station crew. There will be engineers launching and retrieving rockets and satellites. Of course, you will also be there among the throng of paying tourists.

But that is just our first step into space. There will be reason for humans to travel elsewhere, and in this context "elsewhere" can be defined as three places: LEO, GEO and out into space traveling to the Moon or another planet.

To go elsewhere, you need propulsion and a spaceship. Essentially rockets are still needed to jump off of the Elevator, anything from small booster rockets to massive interplanetary craft.

Journeys to LEO

What reasons will we have to journey to Low Earth Orbit (LEO)?

Will it have any practical use? The reason we are there today is mostly about cost. LEO is the cheapest and closest bit of space for us to access. But when you can go to GEO for less cost, what then?

For the sake of the exercise, let's assume that there is still a call for LEO missions. These may be to service the ISS or its successor if such is still there. Some satellites are in LEO orbit and may need servicing. LEO in the 20s will be a crowded place with the advent of CubeSats in their thousands, the main use being to provide internet comms.

LEO will be reached by climbing the ribbon to a height of 700 km or more, and then using booster rockets in an elongated elliptical orbit to reach orbital speed and maneuver to dock with the intended target. For an unmanned mission that is practical enough, but for a manned mission, it involves exposure to high g-force and accelerations, similar to that currently experienced in rocket launches. This will not be a mission for the average tourist but rather for trained astronauts.

What other reason could there be for jumping off in low orbit? How about extreme sports? Think of jumping off at 50 km height wearing a space suit and air-surfing back to the ground!

Journeys to GEO (geostationary earth orbit, 36,500 km away)

In contrast to the LEO experience, a human mission into the GEO orbit is a much more civilized experience. At GEO, all objects already have the correct orbital velocity to maintain their distance from Earth.

It will be possible to launch a spaceship from the GeoStation and for it to slowly circle the Earth, away from the GeoStation. Perhaps there will be continual and regular missions that do the GEO orbit every month, making the space equivalent of house calls for maintenance or delivery purposes, to the hundreds of satellites at GEO.

In earlier books in this series, the focus was on the establishment of the GeoStation at GEO, but there is no reason why this has to be the only space station in this orbit. Arrayed around the Earth in the GEO orbit may be, not just satellites, but other space stations, and that would require GEO orbital missions.

This is possible in theory, but you have to ask for what purpose we would build extra space stations. The GEO Station can become as large as we want it to be, and it is the station with the direct travel link to Earth. But perhaps we will find a need for other space stations that need secrecy or security, or maybe a group of astronomers will want a different position. Who knows? But the basic detail is: if such space stations are built, they can be serviced from the GEO Station.

The other obvious purpose comes when we have more than one space elevator connected to Earth. In earlier books, we postulated four or eight such elevators and each would have its own GEO Station. At that time, the ability to move directly between GEO Stations will be useful.

This will be a trip that can be done by tourists though room will be at a premium. With no noticeable acceleration and g-force to suffer, it will be a continuation of space travel under weightless conditions.

JOURNEYS FROM THE TOP OF THE EARTH RIBBON

Now, at last, we come to the most wondrous journey possible: leaving the Far Far Away Station for the Moon or another planet.

The experience of traveling 100,000 km up the ribbon into space is a wondrous one, and in the 20th century only two-dozen people ever traveled further.

But even at this distance, we are still captive to Earth! We will be rotating around it, still in its gravity well, in its neighborhood.

But the ultimate "jumping off" point is at the end of the ribbon, leaving for other places. Let's assume that we have rejoined you at the end of the ribbon. If we double the cost of your ticket, then you can leave the Elevator and travel to the Moon, and later to Mars and other planets.

Now we are talking about serious space exploration!

For this journey, the level of risk increases. Relative to modern rocket-based space travel, our journey will still be considerably safer. However, once you leave the Elevator ribbon, the risks do increase and your trip will start to feel more like that of a present-day astronaut instead of a plane trip.

At the end of the ribbon, 100,000 km from Earth and 63,500 km from the GEO Station, you can reflect on how remote you are. Literally hanging by a thread, your lifeline to Earth stretches above you, since the Earth is apparently above you. The pressure of gravity is back and hanging below the space station are a number of spacecraft: rocket ships. It looks like an advanced airplane terminal, except for the inky blackness of space.

Boarding a spacecraft

You will be traveling in a much smaller craft, because it will have to be manipulated into and out of orbits and endure changes of velocity and direction. On launch day, you travel down the final kilometer to where your spacecraft is waiting. You will now be suited up into your space suit, as you transfer into the craft and get strapped into your seat.

Safety demonstrations are even more detailed here. You may have undertaken some pilot training: in the event of an emergency everyone on board must be prepared to take on any role. However, this is not an Apollo craft. It is large by comparison with today's spacecraft, and computer controlled, with a backup human controller. If you personally have to take the controls, then everyone is in real trouble!

As the countdown to launch runs down, you are conscious of the dangerous step you are taking. Being suspended on the Elevator ribbon felt fragile, but now you will let go of that umbilical cord and won't even have that for reassurance.

The countdown reaches zero, and there is a small jolt as the spacecraft releases from the ribbon and you drift away from the ribbon. Even so, the apparent loss of gravity means that we are already floating away from Earth and the ribbon. We are on our way to another world.

Now you will be climbing onto a different type of spaceship, one set up for a journey lasting days or months in open space. The spacecraft will be spacious and comfortable, looking somewhat fragile and built for function. One other thing you might notice is that it does have a rocket engine, but you may be surprised at how small it is. The Space Elevator acts like a very large sling and can throw spacecraft to the Moon, Mars, Venus and the asteroids, starting them on their way to the rest of the solar system. Just by letting go of the end of the Elevator at the right time you will be on your way. The rockets are needed on this spacecraft for course correction, additional acceleration and deceleration at the other end.

Adrift in space

After only a few minutes, you and your crew drift out past the final end of the ribbon, where the anchor mass is attached. There goes your last link to home. Ahead is nothing but blackness, punctuated by stars. You thought that the Elevator ride was impressive but now you have a sense of being a real astronaut. Point the craft in the right direction and you could go anywhere. In fact, as time goes by, craft will be doing just that: leaving for every possible destination within our solar system.

There is movement; a major thrust as the computer adjusts the trajectory and rockets fire to build velocity. When that ends, you are traveling to the Moon and weightless again.

Timing is an issue for this journey. Relative to a stationary point on Earth (and hence also relative to the Elevator which stays in a fixed position relative to the Earths' surface) the Moon has an apparent 25-hour orbit, because of the retardation of the Moon in its relative orbit. Your spacecraft is going to have to dock with the Moon elevator, so there is one opportunity every 25 hours to leave the Far Far Away Station, timing it so as to reach the Moon elevator.

Craft returning from the Moon will be doing the opposite calculation: leaving the Moon in time to hook up with the Far Far Away Station once every 25 hours. So, craft will be leaving the Moon at about the same time as your spaceship. Somewhere out there you may be able to spot the returning craft as you pass each other, though you will still be many thousands of kilometers apart.

THE MECHANICS OF THE MOON CONNECTION

Because the gravity on the surface of the Moon is lower than on Earth, it takes less energy for a rocket to leave the Moon. The escape velocity from the Moon is 1.6 km/sec as compared with 9.3 km/sec for Earth.

The distance from Earth to the Moon varies during the Moon's orbit each month, from a distance of some 363,000 km at its closest, or Perigee, to a distance of some 405,000 km at its most distant, or Apogee. That means that the altitude adjustment for Earth-Moon has to shrink and grow by some 42,000 km each month. Also, the Moons' orbit around Earth is highly elliptical with a substantial eccentricity of 5.5%. With this orbital eccentricity and its obliquity to orbit, you have one very complex management problem.

A RECAP

Next idea: if we can't run a ribbon from the Earth to the Moon, what about running a separate ribbon off of the Moon? Then, as the Earth ribbon passes the Moon ribbon, the payload jumps from one ribbon to other, rather like a trapeze artist jumps from one trapeze to another in a circus.

How a Moon ribbon stays up

This raises a number of other critical issues. Firstly, is it possible to have a ribbon on the Moon? The answer is yes, but the orbital dynamics make it a very different venture as compared to our first Earth ribbon.

A reminder: The Earth rotates once every 24 hours. What keeps the ribbon "up" is the rotational speed. The "last" 65,000 km of the ribbon are rotating fast enough to fly away from Earth, were it not balanced by the Earth end of the ribbon.

Distances: the Earth-based Elevator ribbon only stretches a quarter of the way to the Moon.

Now, the Moon rotates on its' own axis only once a month, approximately, or 29.5 days. In other words, the rotational speed of the Moon is only about 3% of that of the Earth. How long would a ribbon have to be, in order to be kept in place by the rotational speed of the Moon? Remember that, for the Earth ribbon, we needed a length of 100,000 km. But the hypothetical Moon ribbon would need to be about 3 million km in length! Regardless of the problems of making a ribbon that length, we can't do it: the Earth is only about 400,000 km away and the ribbon would "hit" the Earth on each rotation.

So, keeping up a ribbon by rotational velocity can't be done for the Moon.

However, the fact that our Earth is 400,000 km away gives us another way of paying out a ribbon:

If we run a ribbon out from the Moon toward the Earth, then there will be a point beyond which the ribbon will behave as though it is hanging "down" towards Earth. The Moon is a big hook and the ribbon hangs down like a long rope!

Another thing: for the first time in our consideration of a Moon ribbon, the universe has done us a favor: the Moon rotates at the same speed that it orbits the Earth, so that the same face of the Moon is always pointed toward the Earth.

If the Moon rotated at any other speed, we would have to dump the whole idea of a Moon ribbon. However, the Moon is almost made for a ribbon to be dropped down into the Earths' gravity well! Next time you stand in your garden and look up at the same old man-in-the-Moon, you can be grateful that you've never seen the far side of the Moon because it's the key to building a Space Elevator on the Moon.

The length of a Moon ribbon

So, let's build our Moon Elevator. How long will it have to be? This is a simple enough equation, balancing the Earths' gravity against the Moons' gravity to find the pivot point, the point at which the gravitational pull of both bodies cancels out. It turns out that this point is well known by another name: L1 or the first Lagrange point where there is a stable gravitational point, or, more literally, a non-gravitational point, since an object at L1 will tend to stay where it is in relation to the Earth and the Moon.

L1 is 80% of the Earth-Moon distance, being closest to the Moon. That puts it some 326,000 km from Earth and 80,000 km from the Moon.

So, if the 80,000 km of ribbon nearest the Moon has a tendency to fall "down" towards the Moon, how much ribbon do we need on the Earth side of L1 to balance it?

Another 220,000 km of ribbon is needed to balance it making

the total ribbon length some 300,000 km (less if we use a counterweight at the end of the ribbon).

This gives an interesting coincidence. The Earth-Moon distance is approx. 400,000 km. If the Earth ribbon is 100,000 km long, and the Moon ribbon is 300,000 km long, then between them they exactly cover the total distance.

However, this is not as useful as it sounds. The two ribbons would be rotating at different speeds and just briefly passing each other every 25 hours.

The trapeze trick

Could you do the trapeze trick and jump from one ribbon to the other as they pass?

The far end of the Earth ribbon, 100,000 km out, will be rotating at a speed of 7 km per second.

The far end of the Moon ribbon, 300,000 km out, with respect to the Earth, will be contributing motion equivalent to one hours' movement per Earth day, which translates to 0.3 km per second: in relative terms the Moon ribbon is almost stationary while the Earth ribbon passes by at 7 km per second, a speed similar to a rocket in flight, or 30 times faster than a jumbo jet. The speed difference has got us again.

A rocket will still have to leave the Earth ribbon and maneuver into an orbit close to the Moon ribbon, but once that rocket has adjusted its speed to that of the Moon, it would be safer and take less energy to dock with a Moon ribbon and ride down to the surface instead of landing like the Apollo modules. It takes less energy to do this adjustment nearer to the Moon, so our rockets will still travel out from the Earth ribbon.

THE MOON BASE
IN DETAIL

Where would be the anchor location or base for the Moon ribbon?

Because of the way it is constructed: hanging "down" towards Earth, dropping into Earths' gravity well, the most stable position would be at the "center" of the Moon as viewed from Earth. The "center" is called the Sinus Medii.

However, the position is not critical, or at least not as critical as it is on Earth. On Earth, the ribbon is kept in place by rotational forces, and so we are concerned to anchor the ribbon within 35 degrees of the Equator at most, preferably as close to the Equator as practical.

On the Moon, though, the fact that the ribbon is hanging down into the Earth's gravity well means that it will "fall" towards Earth, no matter where it is anchored on the Moon. From the point of view of an observer on the Moon, the Earth stays in one place.

So, in practice almost any location on the face of the Moon facing Earth is available to us. However, the angle at which the ribbon meets the surface of the Moon is perpendicular only at the center of the face of the Moon. For any other position, the angle becomes more acute and less useful, so it is likely that, for the first ribbon, we would choose an anchor location within reasonable proximity of the center of the face of the Moon.

Unlike Earth, we do not have weather and atmosphere concerns on the Moon, and because gravity is only one-sixth of the strength of Earth, the lateral strains are weaker. In short, we are given plenty of leeway in deciding where to place it.

Looking at the Moon from Earth, there are many convenient craters close to the "center" of the face of the Moon, such as Alpetragius, the latter being 16 degrees South, 4.5 degrees West and in the original 2006 book, we considered using this as a likely base though, with the benefit of improved surface mapping, the attention has moved to Albategnius, a crater 129 km wide, at 11.2 degrees South, 4.1 degrees East. Albategnius is around 360 km from the Sinus Medii, the "center" of the Moon, so as the first ribbon descends to the Sinus Medii, it only has to be dragged 360 km to its intended destination. The low gravity of the Moon makes this an easy task, compared to the equivalent on Earth.

First, why Alpetragius is attractive. It is a crater 39 km wide, with a mountain in the center.

Photo: Alpetragius marked by arrow, next to Ptolemaeus, Alphonsus and Arzachel.

In this central location of the Moon is a chain of three craters easily seen through a telescope: Ptolemaeus, Alphonsus and Arzachel, familiar to many astronomers from their first views of the Moon. These craters are between 100 km and 300 km wide. Adjacent to both Alphonsus and Arzachel is the much smaller crater of Alpetragius.

Photo: Ranger 9 image of Alphonsus crater (diameter 108 km) from a distance of 442 km, taken about 3 minutes before impact in the upper right portion of the crater. At left is the north-eastern edge of Mare Nubium. The crater adjacent to Alphonsus at the bottom left is the 39 km diameter Alpetragius.

Alpetragius is just 39 km wide. Unlike most craters, it does not simply have a crater floor. Its walls are an amazing 4,000 meters high (4 km) and they descend directly to the foot of a central cone or mountain. The cone itself is 10 km across and 2,000 meters (2 km) high.

We can attach the Space Elevator to the peak of the central cone within Alpetragius. Then we can burrow into the central cone to provide a safe and sheltered terminal, the first Moon Port, and even enlarge it to provide an entire city. A cone 2,000 meters high and 20 km across can be tunneled with caves to provide the space equivalent of a modern city on Earth!

This means that ribbon cars descend down and come to a halt on top of the central cone, in a safe enclosed terminal. This is safer than the traditional idea of rockets powering down towards a Moon colony, where a single rocket accident could destroy it.

The distance from the center of Alpetragius to the center of Albategnius is 300 km, or about 220 km from one crater rim to another. They are close enough to build a covered road or rail between them, though the land between them is rough and littered with small craters.

Albategnius itself contains a number of small craters on its ridge, the largest of which is Klein. Klein is about the same size as Alpetragius and could potentially be a colony base before the larger crater is used. At about 35 km across, it would be easier to put a roof over Klein and create a liveable habitat, from which the colonists can then work on expanding into the whole of Albategnius.

To the north of Albategnius lies the well-known crater of Hipparchus. Interestingly, the two craters are almost joined by a rille, a long, shallow depression, with a high wall on the eastern side, which has the potential to be enclosed and turned into a road or rail link between the craters, once the colony expands.

In the land surrounding Hipparchus, Albategnius, Alphonsus, Arzachel and Alpetragius are many smaller craters. The number of craters, of varying size, gives dozens of candidates for roofing over and creating craters containing an atmosphere and eventually we can imagine the entire region populated with colonies.

As we have explained before, the cost of the new technology is greatest for the first deployments. The first elevator from Earth, the first Elevator on the Moon, the first crater roofed over and filled with air. Once all this is in place, sometime late in the

2030s, we have a colony on the Moon that is viable and the connection to Earth via the Elevator ribbons is cheap and regular. Instead of the penny-pinching approach of current rockets, where every kilogram has to be rationed, we can ship out material by the ton, including the heavy machinery we call earth-movers to shape the lunar surface, to dig into the cones or mountains and create cities within them. This technology changes space exploration from a niche, expensive scheme, into a massive expansion away from Earth into the rest of the solar system.

While Alpetragius is to the apparent left of Alphonsus, viewed from the northern hemisphere on Earth, Albategnius is to the right. Alphonsus itself was the subject of early exploration when Ranger 9 landed there in 1965 and for many decades, the Ranger 9 photos of the area were the best we had. The best photography now available comes from the Lunar Reconnaissance Orbiter (LRO) which has photographed the entire lunar surface in high definition. These photos have recently been stitched together by USGS (United States Geological Survey) into high definition photos of the entire lunar surface and extracts from the USGS mapping are sourced and used in this book.

Earlier it was explained we can drop ribbons to many locations on the lunar surface, akin to the spokes of an umbrella. Additional locations are not identified here, but a question arises: instead of Albategnius, why not just drop the first ribbon to Sinus Medii, the "center" of the Moon and connect it there?

It would certainly be feasible. The lunar surface doesn't have the limitations of choosing locations on Earth, where we are constrained by weather, winds, thunderstorms, and populated areas. On Earth, ocean-based locations are preferred, to enable the ribbon to be moved, perhaps avoiding satellites and debris in LEO. On the Moon, we don't have to contend with weather or objects in orbit (yet), and we don't have the benefit of an ocean, to connect to large ships. Set against this, gravity on the Moon is low, compared to Earth, so it is a manageable process to drag a ribbon to a desired location.

So, when the first ribbon drops to Sinus Medii, what will it find there?

The short answer is, not much. Some of the earliest Moon Landers targeted this area. Surveyor 4 and Surveyor 6 both landed nearby, the principle choice of location being that they landed in the center of a flat, featureless plain. At that time, no-one was sure what the surface was like, whether it was possible to land a craft, so few risks were taken.

Hipparchus is close to Sinus Medii, its loosely defined crater edge being some 70 km from Sinus Medii at its closest point. Tranquillity Base, the landing site of Apollo 11, is about 350 km east of the Sinus Medii, about as far away from it as is Albategnuis to the south-east.

Generally, the Sinus Medii is the demarcation between the cratered areas to the south, and the featureless plains, seas or Mares of the north. The featureless plain containing the Sinus Medii itself does not even have a name but can be regarded as the southern end of Sinus Aestuum, containing the well-known Copernicus crater, or Mare Vaporum which is separated from Sinus Medii by Rima Hyginus, a 200 km long linear rille.

The featureless Mares are best avoided for a ribbon and colony base. The level of electrostatic dust on the lunar surface, or regolith, makes it a difficult spot to operate in, as Apollo 11 discovered.

There are some unnamed craters around Sinus Medii, about 100 km away, but once we decide on the need to drag the first ribbon away from Sinus Medii, then instead of dragging it 100 km why not drag it 360 km to somewhere useful?

By useful, we mean small craters, ideally with connecting rilles, which enable colonies to be constructed within central cones or mountains, craters to be covered with carbon nanotube supported roof membranes, craters filled with air, and connecting roads to nearby colonies. The selected location, Albategnius, offers all of these advantages and over the next century we can imagine the terraforming of this and surrounding craters and hills into a country of its own, supporting millions of people.

WHAT OF THE FAR SIDE OF THE MOON?

News of the far side in 2019 was dominated by the successful Chinese mission to place a lander there, Chang'e 4. The first problem is a communications issue, since spacecraft on the far side are unable to communicate directly with humans back on Earth. So, in addition to Chang'e 4, China put in orbit around the Moon, the Queqiao relay satellite to provide communications between the Chang'e-4 spacecraft and Earth. At this point I note, comms between the near side and the far side are just as difficult.

The Chang'e-4 mission is the first time that a satellite has been sent to an unexplored region on the far side of the Moon, the South Pole-Aitken Basin, a vast impact region in the southern hemisphere. Measuring roughly 2,500 km in diameter and 13 km deep, it is the single-largest impact basin on the Moon and one of the largest in the Solar System.

Chang'e 4 would ideally have been sent to the lunar South Pole in the hope of discovering ice there. But the landing location is (using far side coordinates) 177.5991 degrees east and 45.4446 degrees south.

Consider that location of approx 45 degrees south. That is not at the pole, in fact, it is just half-way between the lunar equator and the south pole, which is indicative of the difficulty in maneuvering a lander away from the equatorial plane. At 45 degrees south, the images returned show a landscape not so different from the near side, hilly terrain with scores of small craters. In general, the far side is far more cratered than the near side. So far, we have not learned a lot from the landing.

Not that this detracts from the incredible achievement by China. The landing of Chang'e 4 is second only to the landing of Apollo 11 in 1969 as the most amazing lunar landing ever. Further, China has a well thought out strategic plan for Moon exploration, in contrast to the USA plans which can seem haphazard and change at the whim of each incoming President.

Which country will build the first space elevator? Which country will return people to the Moon and start building a colony there? The USA may well achieve the first return of people to the Moon but in the long run, I'm inclined to think the Moon will be a Chinese one.

A space elevator can be constructed for the far side of the Moon. It will extend away from the Moon as viewed from Earth (technically it won't be viewed from Earth as it will be "behind" the Moon, but it extends away from Earth) passing through Lagrange Point 2. It will need to be much longer than the ribbon the near side, extending 500,000 km or more from the Moon. Otherwise the same principles apply as for the near side ribbon: the far side ribbon can support multiple spines to connect to several lunar surface locations.

The opposite to the Sinus Medii, the location at 180 degrees longitude, 0 degrees latitude, is heavily cratered and offers any number of potential crater locations with a few kilometers. Among the few to have been named are Coriolis, Krasovsky, Daecalus and Icarus, but they enclose, within their perimeter, many smaller unnamed craters which might do just as well for our purpose. The far side Sinus Medii is marked by a S-shaped rille, the remaining edges of two earlier impact craters, which would be worth exploring.

A particular advantage of a colony on the far side is that the Moon shields it from the comms radiation emanating from Earth. The relative radio silence is good for astronomy - until we go and pollute it with locally sourced radio transmissions on the far side!

If we have ribbons and colonies on the near side and the far side, they will want to be connected. Unfortunately, we cannot use the ribbons to create orbital connections between the two sets of colonies. We revert to using rockets and spaceships to orbit the Moon between one ribbon and another, or we construct road and rail links on the lunar surface to connect the two. As previously mentioned, we can use carbon nanotube "tubes" to create sheltered roadways, isolated from the lunar surface and pumped with air, making surface travel easier.

A key target for all lunar colonies will be the expected deposits of ice, and hence water, at the poles, apparently the south pole in particular. As we cannot construct ribbons directly to the poles, a likely plan is to construct a spine ribbon, running close to the pole, say at 50 degrees latitude, then construct a road or rail link to the water mines of the poles. Transporting water from the poles to the colonies will be a major activity.

This gives the outline to establish genuine large colonies on the Moon, instead of the small ones planned by the current rocket community which will be little more than exploration camps. Existing lunar colony plans are akin to Amundsen's exploration team in Antarctica, while the space elevator based plans are more akin to the discovery and colonization of America. By the 2030s we should be transitioning from one to the other.

A MOON BASE WITH A SPACE ELEVATOR

Moon exploration plans pre-Elevator as at 2020 have been explored in an earlier chapter, but here we consider some broader issues, leading up to how plans change if a space elevator is utilized.

Why go to the Moon again when we could go straight to Mars and build a colony there?

The Mars-first movement has plenty of proponents, not least President Trump, who, I guess, fancies the idea of being the first President to arrange a Mars landing or even land on Mars himself.

But going to the Moon first has advantages. Establishing a colony or base on the Moon or Mars is hard, very hard. The Moon is closer, much closer than Mars, so it makes a good place to test our space faring abilities and learn from them. Mars can be so far away, even on the other side of the Sun every second year, that a Mars mission is truly on its own. It isn't easy to send out needed help from Earth. In contrast, once we have regular travel to the Moon set up, help, provided by additional spaceships from Earth, can get there in a few days.

In fact, much can be done by remote control on the Moon, from Earth, preparing a colony for humans, so minimizing the risks that accompany the initial establishment phase. Radio signals take just 1.3 seconds to reach the Moon, which is close enough to instant to be useful. With the aid of additional satellites, a colony on the Moon can be linked into the Internet and global comms.

But signals to Mars take longer. At its closest to us, a signal to

Mars still takes over 3 minutes, at its furthest, over 22 minutes, so remote control is more difficult. We do it now, with the rovers on Mars, but they move in a meticulous, slow manner, to avoid accidents.

SURFACE TEMPERATURE ISSUES

So, we land on the Moon, erect some kind of habitat and live in it, right? Not so fast. The lunar surface (and that of Mars) is a place of extremes and dangers, starting with the temperature.

The Moon has no atmosphere to speak of. For practical purposes, it is in a vacuum, not much different from space in general. An atmosphere, such as we have on Earth, is valuable, not just for breathing, but for air pressure, and for spreading heat around.

The lunar surface experiences huge temperature extremes. Recall the lunar "day" lasts for around a month. For approximately two weeks it is bathed in sunshine with temperatures reaching 127°C. Then, in the two week "night" temperatures plummet to -173°C.

We don't quite get those temperatures on Earth but imagine trying to live in the middle of the sandy Arabian desert in summer, then getting picked up and dropped by the South Pole for winter, only worse.

But the Apollo astronauts lived on the Moon for a few days each trip and survived, didn't they?

Yes, they did, but they didn't live there for an entire month. There were a couple of tricks used, to keep them alive.

Looking at a map of the Apollo moon landing sites, well, they look random, but they were anything but. Apart from choosing a smooth landing site, the missions were timed to arrive the equivalent of ten in the morning, in an Earth day. The sun had been warming the surface for a few Earth days, so the lunar surface temperature had risen from that chilly -173°C to be above freezing, close to the ambient Earth temperature. This is the "Goldilocks" time on the Moon, and for a few of our days, the temperature is mild.

So, the astronauts landed when and where the lunar surface was warm enough not to be frozen and left before it got too warm. Had they intended to stay for an entire month, an entire lunar "day", they could not have managed in the Apollo module. They would have baked to death long before night, when they would have frozen to death.

Then there were the spacesuits. Imagine standing on the lunar surface without a spacesuit on (we'll let you wear jeans and T shirt). Forgetting about the vacuum for a moment, the side of you facing the sun would be hot, while your back would be freezing.

I remember something of the same as a child, one chilly winter in England, like we don't get any more due to climate change. It was around -20°C at night. We had a big bonfire burning to keep us warm. I stood near it to keep warm, like everyone. But my front, facing the bonfire, would be hot from the radiated heat, while my back was freezing, so I had to rotate to keep all of me warm.

Translate that to the lunar surface. When you walk in the sunshine, you are getting the full blast of the sun. But drop into a crater in shadow and you are immediately in freezing cold. There is no average temperature.

Temperature extremes are a killer on the lunar surface, literally. To live on the surface permanently requires a large and insulated habitat, the sort of thing that probably won't be available to the first return missions.

Even getting into the shadow of a crater doesn't help. On Earth, the atmosphere means that in twilight conditions at sunrise, sunset, the temperature gets averaged out. On the Moon, you are either in sunlight, and hot, or in shadow, and cold, nothing in between.

THE PROBLEM OF METEORITES

As if temperature extremes weren't enough, there is the problem of being hit by meteorites while on the lunar surface.

On Earth, meteorites burn up in our atmosphere. There are so many we get regular displays of these to watch, sometimes called "shooting stars". Larger meteorites do reach the ground and we call these meteors.

But the Moon doesn't have an atmosphere so even small meteorites don't burn up - they carry on until they strike the surface. This means it is possible to be hit by such rocks, at best, puncturing a spacesuit and creating an air pressure emergency, at worst killing you. Habitats can be punctured by meteorites with the same problems.

Being hit by a meteorite on the Moon is not a common event - the Apollo astronauts survived without being hit. But stay on the Moon permanently - sooner or later it will happen.

This is another good reason to build a habitat underground, where the surface can provide protection.

BUILDING A FIRST HABITAT

To build a sustainable colony on the Moon, the surface is to be avoided - the colony needs to be underground. There are ideal places for underground colonies in the center of many craters, which have a large cone or mountain. Think of them as a solid office block, just waiting to be tunneled into. We can create the sort of space humans are used to. Just like our office blocks and apartments, we build a city inside the mountain, with carved out rooms connected by tunnels. The prime rooms on the outside edge can have insulated windows looking out onto the crater.

When we have a space elevator on the Moon, it can be connected to the peak of the mountain, providing transport on and off the Moon.

Eventually, we can cover the entire crater with a carbon nanotube supported membrane, stretching out like and umbrella. Pump air into the roofed crater, and finally we have a space that is hospitable for humans.

Can we do all this without a space elevator?

Well, anything is possible, but if our nascent colony is supported by rockets only, there are limits on capacity. If a spaceship is arriving by rocket, on a monthly basis, which seems the best we can do at present, then the supplies it carries will be limited to air, food and water to maintain life, with tools and fuel. So, building and expanding the habitat will be a slow process.

Secondly, to start excavating into a mountain, to create an underground colony, we want heavy digging equipment, the sort of earth-moving equipment used in mines. These are large and heavy, so that many are too large and heavy to be taken by rocket to the Moon. Even if we take them in bits and assemble them on the Moon, they need massive amounts of fuel, air and spare parts. It can be done, but it will be challenging and slow.

In the scenario without a space elevator to transport heavy equipment on a daily basis, we'd be looking at smaller excavators and associated equipment. Scaling down, in this manner, means extending the time taken to build a habitable space inside a mountain or underground elsewhere.

BUILDING A FIRST COLONY BY REMOTE CONTROL

When we are ready to build a first colony on the Moon, there are advantages to doing it by remote control and with AI, rather than sending humans at first.

Of course, there will be human missions, initially. After so long an absence from the Moon, we'll want to send humans there to plant a flag, show the surface in high-def TV, put an end to "they didn't go" conspiracy theories and show we are back on the lunar surface.

But having done that, humans are a nuisance on the Moon. If they stay for long, they need a temperature-proof habitat from the start, and they need regular supplies of air, food and water.

Better to send spaceships packed with cargo to the landing site. Then, without the worry of maintaining humans, the machines can assemble a larger habitat, perhaps dig into the ground to make an underground one, build stores for fuel, water, air and food.

Radio signals from Earth take just 1.3 seconds to get there, so it's quite feasible to remote control machinery from Earth. Or, by the 2030s, advances in AI probably mean many machines can operate in autonomous mode. This is in contrast to the equivalent exercise on Mars, where the lengthy comms times mean the building process cannot be monitored in near real time.

By the time the next human missions return to the Moon,

there will be an entire habitat ready and waiting.

WHAT CHANGES IF WE WAIT FOR A SPACE ELEVATOR?

Some things don't change: dealing with the heat of day and the cold of night. Our fledgling colony in Albategnius or wherever is either bathed in sunlight all the time, with heat and radiation, or in the middle of the night, with cold and the image of Earth in the sky for company. This makes the idea of burrowing into the central cone a logical one as the site of the first colony.

Some things do change: shipping Caterpillar® earth-moving equipment and similar to the Moon for example. It's the sheer quantity of materials that can be shipped that changes Moon exploration.

When you can have several deliveries a day, Moon exploration and bases become viable. We can also service a number of bases, instead of just having one.

Where we would like to go? The South Pole for water, the North Pole for perpetual daylight, the far side for astronomy?

Digging bases into the ground, and covering small craters, for living space, will be viable with the right equipment.

A City in the Central Cone of Albategnius

What will they find there? Well, after a bit of construction work, an entire city. It will become possible to burrow into the central cone and turn it into an entire habitat, built around a network of tunnels within the cone. The first Moon base could well be built inside the Albategnius cone, where it can be 10 km wide and with 2,000 or more meters of height to play with, it will be an office block to dwarf any office block on Earth!

COVERING ALBATEGNIUS WITH A ROOF

But that is not all. We have an incredibly strong material in carbon nanotube. We can erect a netting of carbon nanotube running from the crater rim of Albategnius to the top of the central cone, and this netting could support an airtight membrane: we can enclose the entire crater and fill it with air! This will give us our first true Moon colony location, and a model for replicating Moon bases in other locations.

In fact, thanks to our Space Elevators, we can build something very large, city-size, because of the ease with which we can freight items to the Moon.

The present rocket-based plans will of necessity result in a small Moon base, limited by the relatively small launch capacity. But with Space Elevators in place, we can have many cars a day landing on the Moon, each of which can carry 13 tons Earth weight. It becomes quite practical to deliver hundreds of people to the Moon, along with the supplies necessary for their survival. The capacity of each delivery means that large earthmoving vehicles, tractors, and the like, can be taken to the Moon in their entirety.

In their first Moon base, hollowed out inside the central mountain cone, the first people on the Moon will be something more than mere visitors. They will be settlers establishing a real city and colony, protected from radiation by the bulk of the central cone above their heads.

An atmosphere in the crater

The imaginative use of a carbon-nanotube netting, covering the entire crater and containing an atmosphere means that people can move around without a spacesuit. It doesn't take much imagination to realize that, with Space Elevators and carbon nanotube fabric, we can create a liveable environment far superior to anything that is dependent solely on rocket ships for supplies.

As to what we can do with a crater having an atmosphere, well that depends on sunlight and water. Until we learn how to synthesize water on the Moon, we will still be dependent on shipping water in from Earth or extracting it from the lunar poles. At least, with the Space Elevator, the cost will not be prohibitive, but water will still be an expensive commodity on the Moon, so visions of filling up a crater to make a lake are probably not likely at first.

Growing crops will be an obvious usage of the crater area, but we have problems with sunlight. The crater moves into full sun for about 14 Earth-days straight, followed by 14 Earth-days of night. This implies great temperature differences too. However, with an atmosphere and a roof, we can develop technology to offset these disadvantages. We can provide artificial light to assist with crop growth in the night as well as heating.

Power for light and heating will have to come from power plants. It may well be that these are nuclear power plants, located well within a crater wall, or even another crater, for safety, and providing the power necessary to maintain a useful temperature and ambiance in the crater, as well as powering the operations of the Space Elevator ribbon.

Spiders web

As is the case for Earth, we would like to build more than one Space Elevator ribbon to cover the eventuality of a ribbon break. Unlike the Earth, we can't spread the ribbons out around the Moon and have them pointing in different directions like the Earth ribbons. Basically, all Moon ribbons dropping down to Earth will follow a common trajectory, meeting in space at L1, although they could be tethered to different spots on the Moon. Given the low gravity environment, having two elevator ribbons running adjacent to each other is not a problem. It just becomes a management issue in ensuring that they don't get tangled, and that ribbon cars don't hit each other.

TRAVEL ON THE MOON

What about travel between locations on the Moon?

There is another interesting application for carbon-nanotube ribbons. Travel on the Moon surface will be a slow and dusty affair until we lay roads. Moon dust has a static attraction and can work its way into moving parts.

Airplanes are unusable as there is no atmosphere to hold a plane up. We can imagine using rockets to hop around the Moon but that is relatively expensive and carries safety risks.

However, with a ribbon hanging down towards Earth, we could attach secondary ribbons, spreading out from the main ribbon somewhat like the spokes of an umbrella. Each secondary ribbon can be attached to a separate location. With some switching controls, we can then have ribbon cars climbing up one secondary ribbon and switching to descend down another ribbon. This is a safer mode of transport than rockets and probably much safer.

This works because (a) there is no atmosphere on the Moon and (b) the gravity is low.

It has been suggested that the same idea be used on Earth to provide transport between locations. However, on Earth the complications of atmosphere, gravity and weather make it impractical. Plus, on Earth we have airplanes, as a viable alternative, so there is no economic imperative to do it.

Traveling elsewhere: The slingshot effect

The Moon is but one step in the journey into space. How do we leave the ribbon and travel elsewhere?

Momentum at the end of the ribbon

Momentum is a useful thing. Any spaceship at the end of a rotating ribbon has the same velocity and momentum. On letting go of the ribbon, the spaceship would move off in a straight line (admittedly adjusted for the gravitational pull of all nearby large bodies) at a tangent to the ribbon and maintain its speed. That can be most useful because our spaceship doesn't have to start moving from a standing start, unlike a rocket ship on Earth or the Moon.

The ribbon dance

All movements up and down the ribbon, and on or off it, affect the momentum of the ribbon and the balance of the whole system.

The spaceship gets a momentum boost by traveling to the top of the ribbon. But the ribbon also loses that momentum when the spaceship departs. What effect does that have on the ribbon system?

A load, e.g. a car traveling up the ribbon from Earth, has a low mass relative to the ribbon system as a whole. Having only the rotational velocity of the Earth's surface, it will tend to act as a "drag" on the ribbon as it ascends, pulling it "backwards". If it were too heavy relative to the ribbon system, eventually the ribbon would curl "backwards" and wrap itself around the Earth, with the whole thing falling down to the surface. However, because the ribbon car mass is less than 1% of the total system mass, it will eventually adopt the rotational velocity of the ribbon. As a one-off, it speeds up but does it by causing the Earth to slow down in its rotation marginally.

However, in practice, once up and running there will be loads going both up and down the ribbon. A load coming down the ribbon will experience the opposite effect (and also speed up the Earth's rotation again to balance the other cars).

Thus, a ribbon experiencing perfectly balanced loading, with ribbon cars going up and down at the same time, would cause no net movements. It is as though the ribbon cars descending pass their momentum over to the ribbon cars ascending and it all stays in balance.

Of course, the system won't always be in balance. Especially at first, there is likely to be an excess of upward movements as we send people and goods into space. Those same people have to come back sometime so their mass will soon balance. Only later will we experience significant inbound movements of goods.

So as at 2020, where are we at with Moon exploration?
The outlook is for:

- Exploratory missions, landing rovers for research purposes, during the 2020s.
- A crewed landing by NASA scheduled for 2024 is more likely to be 2028. Not long after that, crewed landings by China and perhaps Russia will follow.
- For many years after the first crewed landing, repeat missions will be brief, perhaps annual, events, aimed at exploration and sample return.
- The first permanent or semi-permanent lunar colony is likely to happen in the early 2030s, perhaps by the USA, but more likely by China. It will be preceded by autonomous robots pre-building a base, AI controlled, ready for later occupation by humans.
- Sometime during the 2030s, the technology for a lunar space elevator will be ready for deployment, which will be a game-changer, opening up lunar bases and exploration by a timetable of high frequency space flights via space elevators from Earth.
- By 2040 we can expect to see a number of lunar bases.

A Space Elevator on the Moon

Book 5 of the Space Elevator 2020 series

Published 2020

Linda J. Phillips

A Space Elevator on the Moon

Publisher contact: info@21stcentury.space
FaceBook www.facebook.com/lindyjaniceAuthor
Twitter @_lindaphillips
Amazon author page https://www.amazon.com/author/lindajanicephillips
Web 21stcentury.space
Web links utilized in this publication were correct at the time of writing, but they can change over time.

Published by Linda Phillips